Cambridge Primary

Science

Second Edition

Workbook 6

Series editors:
Judith Amery
Rosemary Feasey

Boost

HODDER
EDUCATION
AN HACHETTE UK COMPANY

Cambridge International copyright material in this publication is reproduced under licence and remains the intellectual property of Cambridge Assessment International Education.

Registered Cambridge International Schools benefit from high-quality programmes, assessments and a wide range of support so that teachers can effectively deliver Cambridge Primary. Visit www.cambridgeinternational.org/primary to find out more.

Acknowledgements

The Publishers would like to thank the following for permission to reproduce copyright material.

Photo acknowledgements
p. 13 *tl, cr,* **p. 17** *tl, cr,* **p. 25** *tl, cr,* **p. 35** *tl, cr,* **p. 43** *tl, cr,* **p. 53** *tl, cr,* **p. 61** *tl, cr,* **p. 71** *tl, cr,* **p. 80** *tl, cr,* © Stocker Team/Adobe Stock Photo.

t = top, *b* = bottom, *l* = left, *r* = right, *c* = centre

Every effort has been made to trace all copyright holders, but if any have been inadvertently overlooked, the Publishers will be pleased to make the necessary arrangements at the first opportunity.

Hachette UK's policy is to use papers that are natural, renewable and recyclable products and made from wood grown in well-managed forests and other controlled sources. The logging and manufacturing processes are expected to conform to the environmental regulations of the country of origin.

Orders: please contact Hachette UK Distribution, Hely Hutchinson Centre, Milton Road, Didcot, Oxfordshire, OX11 7HH. Telephone: +44 (0)1235 827827. Email education@hachette.co.uk Lines are open from 9 a.m. to 5 p.m., Monday to Friday. You can also order through our website: www.hoddereducation.com

© Rosemary Feasey, Deborah Herridge, Helen Lewis, Tara Liveseley, Andrea Mapplebeck and Hellen Ward 2021

First published in 2017

This edition published in 2021 by

Hodder Education,

An Hachette UK Company

Carmelite House

50 Victoria Embankment

London EC4Y 0DZ

www.hoddereducation.com

Impression number   10 9 8 7 6 5 4 3 2

Year   2025 2024 2023 2022 2021

Cover illustration by Lisa Hunt, The Bright Agency

Illustrations by Jeanne du Plessis, Natalie and Tamsin Hinrichsen, Stephan Theron, Vian Oelofsen

Typeset in FS Albert 12/14 by IO Publishing CC

Printed in the United Kingdom

A catalogue record for this title is available from the British Library.

ISBN: 9781398301559

Contents

Body organs

1 Write the name of each organ in the correct column of the table.

heart   nose   kidneys   intestines   stomach

gallbladder   brain   lungs   skin   ears

Internal organs	External organs

2 Choose an internal organ from question 1. Draw and label a diagram of it. Add a title. Write a caption to describe its function.

Title: _____

Labelled diagram:

Caption: _____

Heart and circulation

1 Tick true (✔) or false (✗) for each statement below.

	Statement	True	False
a	Your heart is a muscle.		
b	Your heart is in your chest, slightly to the right of the centre.		
c	Your heart is about as big as your head.		
d	Your heart is always beating.		
e	Your heart pumps blood around your body in a three-stage process.		

2 The diagram below shows the heart and circulatory system. Label the diagram. Use these words:

(heart) (lungs) (artery) (vein) (rest of body)

One has been done for you.

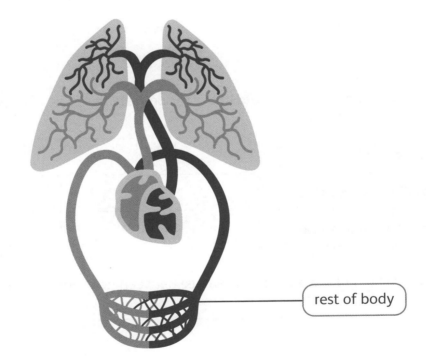

rest of body

3 Describe two ways in which your heart is different to your other muscles.

- _____
- _____

Heart rate and exercise

 1 Some Class 6 learners measured their heart rates before and during exercise. They collected these two sets of data:

Zou   140
Nadia   115
Tyrone 125
Claudia 110
Lionel 120
Sergio 120

Zou   90
Nadia   80
Tyrone 85
Claudia 80
Lionel 75
Sergio 70

a Which set of data do you think shows heart rate during exercise? Explain.

_____

_____

2 **a** Use the two sets of data above to complete this table of results.

Name of learner						
Heart rate before exercise						
Heart rate during exercise						

b Who had the greatest difference between heart rate before and during exercise?

_____

3 Explain why heart rate is higher during exercise.

_____

_____

_____

Process of breathing

1 Complete the table below about the process of breathing. Use these words and phrases:

carbon dioxide is breathed out   move upwards   fill up with air

flattens and moves downwards   exhalation

Stage of breathing	What your diaphragm does	What your ribs do	What your lungs do	What happens inside your lungs
inhalation		lift upwards and outwards		oxygen passes into the bloodstream
	moves upwards		air is pushed out of them	

2 a Draw and label a diagram to show what happens during inhalation.

b Draw and label a diagram to show what happens during exhalation.

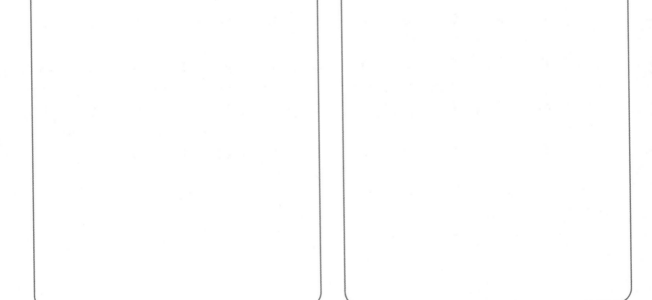

Infectious diseases word search

1 Find and circle these words in the word search below.

infectious disease microbes microscopic infected bacteria

virus fungi parasite cold malaria cough hygiene transmission

The words may be written in any of these directions:

N	I	N	F	P	A	R	A	S	I	T	E
D	O	M	I	S	M	I	C	I	N	F	E
B	C	I	N	F	E	C	T	I	O	U	S
A	O	C	S	U	R	I	V	H	A	N	E
A	U	R	V	S	D	F	U	Y	I	G	B
I	G	O	I	P	I	C	O	G	R	I	O
R	H	S	R	A	S	M	I	I	E	N	R
A	M	C	C	R	E	F	S	E	T	F	C
L	A	O	O	H	A	D	U	N	C	E	I
A	L	P	U	Y	S	L	R	E	A	L	M
M	A	I	N	G	E	O	I	A	B	R	I
F	U	C	D	E	T	C	E	F	N	I	T

Microbes

1 Fill in the blank spaces in this paragraph, using only the words *viruses* and *bacteria*.

There are different types of microbes on Earth. Two of the main types are _____ and _____. _____ can live in almost every environment, including in and on the human body. Most _____ are harmless. Some _____ can be beneficial to humans. They can help us digest food and fight disease-causing microbes. Less than 1% of _____ cause diseases in humans. Most _____ do cause disease and can be more harmful to humans.

Both _____ and _____ are too small to be seen by the naked eye, and _____ are much smaller than _____. _____ can survive without a host, but _____ can only survive once they get into the cells of a host. Both _____ and _____ can be spread in a number of different ways. A person who has been infected by harmful _____ can be treated with antibiotics. However, a person who has been infected by _____ cannot.

2 **a** Write three questions that can only be answered with the word *bacteria*.

_____

_____

_____

_____

_____

b Write three questions that can only be answered with the words *virus* or *viruses*.

_____

_____

_____

_____

_____

Transcription of diseases

1 Diseases can be spread in different ways. For each picture below, state: how the disease could be transmitted and suggest a way it could be prevented or stopped.

Use these words:

(air) (touch) (water)

(food) (animals)

a

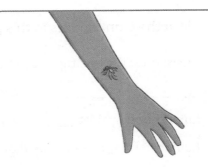

The disease is being transmitted by an **animal**. The transmission could be prevented by wearing an insect repellent.

b

_____

_____

_____

c

_____

_____

_____

d

_____

_____

_____

e

_____

_____

_____

Hand hygiene

Some Class 6 learners investigated the difference that washing hands with soap makes to the spread of germs. They took five slices of bread from the same loaf. The table below shows what the learners did to each slice. They then placed the slices in sealed plastic bags. After a month, they measured how much of the surface area of each slice was covered in mould.

Bread	Surface area of bread covered in mould after one month
Untouched	5 %
Touched with unwashed hands	60 %
Touched with hands washed with soap and water	10 %
Touched with hands cleaned with sanitiser	25 %
Wiped on the keyboard of the class computer	90 %

a   Draw a bar chart to show the data in the table above. Remember to give your bar chart a title and to label both axes.

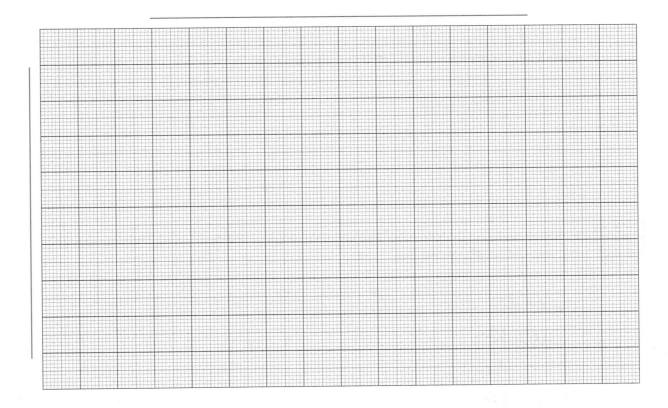

b   Use the information collected to advise the learners where to improve the hygiene in their classroom. How could they do this?

_____

_____

Human defence mechanisms

1 Humans have different defence mechanisms to help prevent infectious diseases harming the body. On the table below, write which one or more of these three defence mechanisms is being used to protect the human body from potential infections:

skin   stomach acid   mucus

One has been done for you. Add three more scenarios of your own at the bottom of the table.

Scenario	Body's defence mechanism
Sand and dust blowing into your eyes	mucus (tears)
Eating food that has been dropped on the floor	
Picking up a ball off the ground	
Coughing and covering your mouth with your hand	
Breathing harmful bacteria in through your nose	

2 Produce an acrostic for either the words *mucus* or *stomach acid*. Here is an example of an acrostic for the word *skin*:

defend**S**
attac**K**s
In
huma**N**s

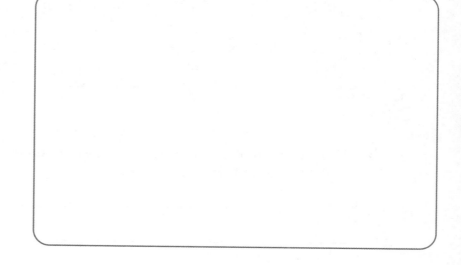

Unit 1 Systems and diseases

Self-check

See how much you know!

 I can do this.

 I can do this, but I need to keep trying.

 I can't do this yet.

What can I do?			
1	I can name the different parts of the human circulatory system and explain its function.			
2	I can describe how the heart pumps blood through arteries, capillaries and veins.			
3	I can name the different parts of the human respiratory system and explain that it gets oxygen into the blood.			
4	I can name vertebrates that have similar circulatory systems or similar respiratory systems to humans.			
5	I can name the different microbes that can infect a person.			
6	I can name three ways diseases spread.			
7	I can describe how good hygiene can control the spread of diseases.			
8	I can describe at least three natural defence mechanisms humans have to prevent infection.			
9	I can link ideas about the different ways diseases spread to good hygiene standards and human defence mechanisms.			

I need more help with:

_____

_____

_____

Reproduction word search

1   Find and circle these words in the word search below.

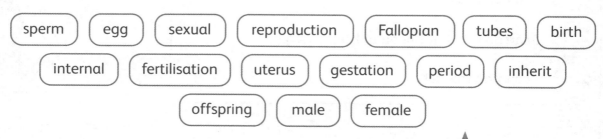

sperm   egg   sexual   reproduction   Fallopian   tubes   birth

internal   fertilisation   uterus   gestation   period   inherit

offspring   male   female

The words may be written in any of these directions:

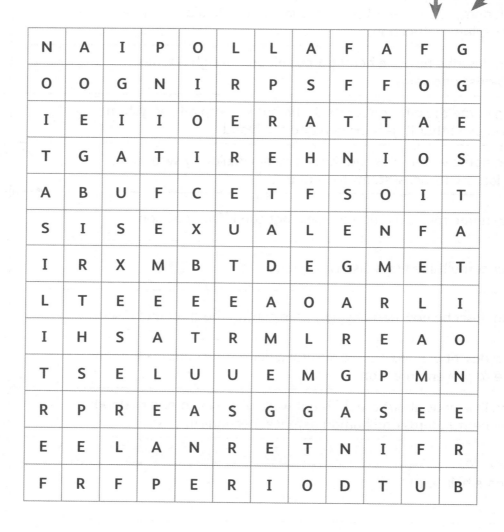

N	A	I	P	O	L	L	A	F	A	F	G
O	O	G	N	I	R	P	S	F	F	O	G
I	E	I	I	O	E	R	A	T	T	A	E
T	G	A	T	I	R	E	H	N	I	O	S
A	B	U	F	C	E	T	F	S	O	I	T
S	I	S	E	X	U	A	L	E	N	F	A
I	R	X	M	B	T	D	E	G	M	E	T
L	T	E	E	E	E	A	O	A	R	L	I
I	H	S	A	T	R	M	L	R	E	A	O
T	S	E	L	U	U	E	M	G	P	M	N
R	P	R	E	A	S	G	G	A	S	E	E
E	E	L	A	N	R	E	T	N	I	F	R
F	R	F	P	E	R	I	O	D	T	U	B

Puberty mind map

1 Create a mind map to show what you know about puberty. Here are some words to use:

(growth) (male) (female) (menstruation) (hair) (voice)

puberty

Keep adding to this mind map throughout the unit. When you learn something new, use different-coloured pencils or pens to add to it.

Comparing reproduction in different animals

Below are the names of the male, female and offspring of different animals.

(lion cub) (lioness) (lion)   (tadpole) (female frog) (male frog)

(baby whale) (female whale) (male whale)   (caterpillar) (female butterfly) (male butterfly)

(female turtle) (male turtle) (baby turtle)   (baby alligator) (female alligator) (male alligator)

1 What do these animals have in common with human reproduction?
Write down three ideas.

- _____

  _____

- _____

  _____

- _____

  _____

2 Explain why it is important for animals, including humans, to reproduce.

_____

_____

3 Use the animals listed above to create a group of three animals that have something in common. Write what they have in common. Then decide which of the three is the 'odd one out'. Think about what is different about how this animal reproduces, compared to the other two.
One has been done for you.

a The group is: **alligator**, **whale**, and **turtle**. They all **live in water**.

   The odd one out is **whale** because **it gives birth to live young**, but the turtle and alligator lays eggs.

b The group is: _____, _____, _____.

   They all _____.

   The odd one out is _____ because _____.

c Share your group of three animals with a partner. See if they came up with the same or different 'odd one out' to you.

Unit 2 — Human reproduction

Self-check

 I can do this.

 I can do this, but I need to keep trying.

 I can't do this yet.

See how much you know!

What can I do?			
1 I can name the parts of the male reproductive system.			
2 I can name the parts of the female reproductive system.			
3 I can compare reproduction in different animals.			
4 I can explain why it is important for all animals to reproduce.			
5 I can describe the different life stages that humans go through.			
6 I can describe the physical changes that take place during puberty in males and females.			

I need more help with:

_____

_____

_____

A seashore food web

1 Write the names of some plants and animals that live in a seashore habitat. Make notes about their feeding relationships.

You will need to do some research.

2 Draw a food web to show the feeding relationships from question 1.

Food webs

1 Some Class 6 learners produced food webs to show how energy moves through an ecosystem. Can you spot errors in their work? Write notes below each food web to tell them what they have done incorrectly and how to correct their food webs.

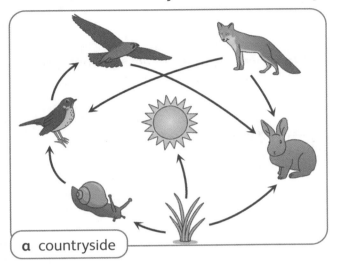

a countryside

_____

_____

_____

_____

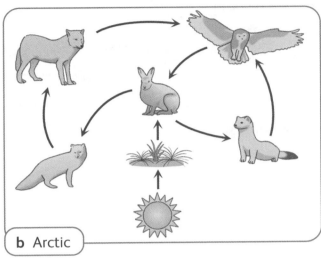

b Arctic

_____

_____

_____

_____

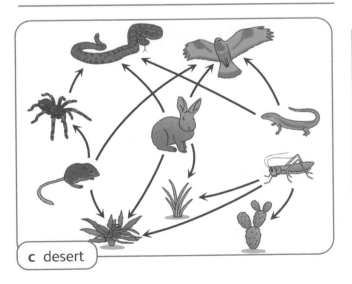

c desert

_____

_____

_____

_____

d rainforest

_____

_____

_____

_____

Food chains

1 Predict what might happen to the organisms in the food chain when there is a change in the ecosystem. Explain your prediction.

a

Food chain	Change
leaf → caterpillar → frog → snake	Warm, wet weather speeds up plant growth, and leads to more leaves than usual.
Prediction	**Explanation**

b

Food chain	Change
fig → tapir → boa constrictor → harpy eagle	A disease that affects fig trees leads to fewer figs.
Prediction	**Explanation**

c

Food chain	Change
shrimp · swordfish phytoplankton · bluefish	Humans catch all the swordfish.
Prediction	**Explanation**

Endangered species fact files

 a Choose two animal species that are endangered. You could choose two from these, or any other two:

mountain gorilla   blue whale   snow leopard

black rhino   red panda   orangutan

b Fill in the fact file about each endangered animal. Do research to help you.

c Draw or stick a picture of the animal onto the fact file.

Common name:

Scientific name:

Habitat(s):

Diet:

Average life span: Size:

Weight: Number in the wild:

Threats to the ecosystem:

Common name:

Scientific name:

Habitat(s):

Diet:

Average life span: Size:

Weight: Number in the wild:

Threats to the ecosystem:

Waste survey

Some Class 6 learners carried out a waste survey at their school. Here are their results:

Survey questions	Answers
How many classrooms have a paper recycling bin?	10 out of 10
Does every classroom collect junk materials for modelling?	Yes
Does the school have a compost bin for fruit and vegetable waste?	No
Does the school encourage learners to use refillable drink bottles?	Yes
How many litter bins are in the playground?	1
Do the playground litter bins divide litter into recyclable and non-recyclable items?	No
Does the school have a policy of mending broken items wherever possible?	Yes
Does the school encourage learners to bring only unpacked snacks to school?	No

 Write an email to the head teacher of the school. Suggest actions that the school could take to improve the way it manages waste.

New Message	− ↗ X
To	**Cc Bcc**
Subject	

_____

_____

_____

_____

_____

_____

_____

_____

_____

_____

Send

Sort that waste

1 Into which bin should you put each waste item? Write the name of each waste item on the correct bin.

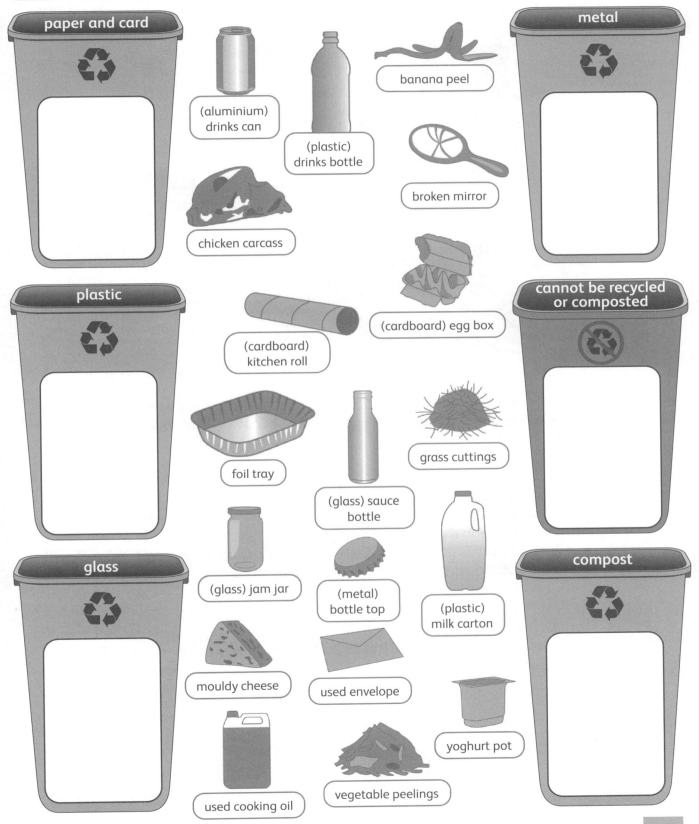

paper and card

metal

(aluminium) drinks can

(plastic) drinks bottle

banana peel

broken mirror

chicken carcass

plastic

cannot be recycled or composted

(cardboard) kitchen roll

(cardboard) egg box

foil tray

grass cuttings

(glass) sauce bottle

glass

compost

(glass) jam jar

(metal) bottle top

(plastic) milk carton

mouldy cheese

used envelope

yoghurt pot

used cooking oil

vegetable peelings

23

Toxic accumulation

DDT is a chemical that has been used by farmers as an insecticide. The rain washes DDT into rivers and then into the ocean. DDT then contaminates water ecosystems. Fish take up DDT and store the chemical as DDE in their bodies. Animals cannot break down DDE.

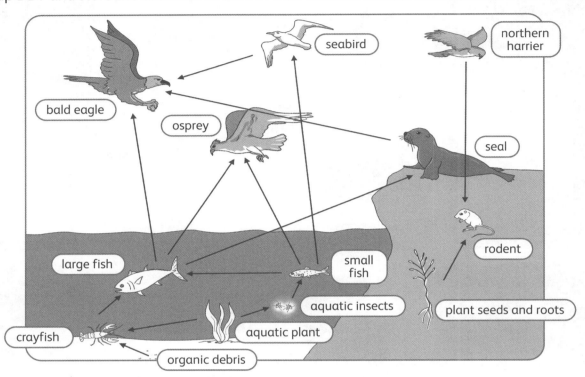

a   An osprey eats 400 g of fish per day. The fish tissue that the osprey eats has an average DDE concentration of 0.1 micrograms per gram. How much DDE is the osprey eating in one day?

_____

b   A bald eagle also eats 400 g of food per day. It eats fish, as well as seal carcasses that have washed up on the beach. A seal eats fish-eating fish with 1.0 microgram per gram of DDE in their tissue. The DDE builds up in the seal's body so that the seal has 2.0 micrograms per gram DDE in its tissue. If the bald eagle eats 200 g of seal, how much DDE does the bald eagle consume in one day?

_____

c   Take all the above factors into account. Rank the following animals in order, from most likely to have built up DDE in its body to least likely to have built up DDE. Explain your reasons.

| bald eagles | osprey | seals |

_____

_____

_____

Unit 3   Ecosystems

Self-check

See how much you know!

 I can do this.

 I can do this, but I need to keep trying.

 I can't do this yet.

What can I do?			
1 I can explain how a food web is different to a food chain.			
2 I can state where all animals on Earth get their energy from.			
3 I can explain how energy is passed through a food chain.			
4 I can draw a diagram to represent how energy is passed through a food chain.			
5 I can explain what a toxic substance is.			
6 I can describe at least four different ways in which human activities add toxic substances to an ecosystem.			
7 I can explain how toxic substances move through a food chain/web.			
8 I can explain why toxic substances are more harmful to predators at the end of a food chain.			

I need more help with:

_____

_____

_____

Changes to materials

1

   a   Draw and label a diagram to show a reversible change.

   b   Draw and label a diagram to show an irreversible change.

2   Look at the diagrams you drew in question 1. Decide which of the materials you could change back again. Draw and label a diagram to show how you would do it.

Investigate melting

1 Some Class 6 learners investigated how quickly ice would melt in different locations around their school. They left a block of ice to melt and, every 2 minutes, measured its mass. Here is their table of results.

Location	Mass of ice at beginning of test	Mass of ice after 2 minutes	Mass of ice after 4 minutes	Mass of ice after 6 minutes	Mass of ice after 8 minutes
A	16 g	12.5 g	9 g	5.5 g	2 g
B	16 g	12 g	8 g	4 g	0 g
C	16 g	13.5 g	11 g	8.5 g	6 g
D	16 g	13 g	10 g	7 g	4 g

Plot the data from the table on a line graph. Use a different colour to show the data for each location. One set of data, for location A, has been plotted for you.

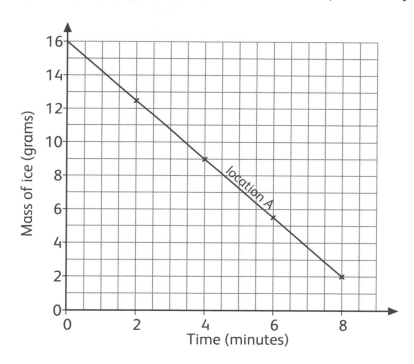

2 a Order the locations according to their temperature, from highest to lowest.

_____

b Explain how you know which location had the highest temperature.

_____

_____

Freezing different liquids

1 Some Class 6 learners set up an investigation to find out how the thickness of a liquid affects how fast it freezes. This is what they used:

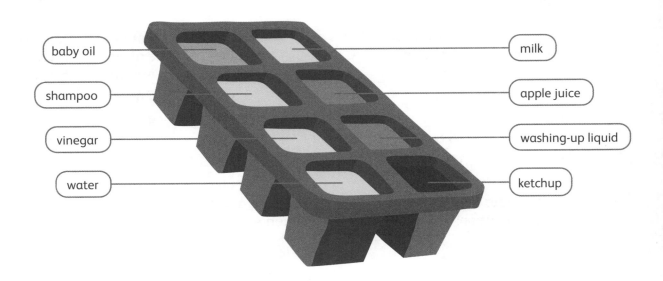

baby oil

shampoo

vinegar

water

milk

apple juice

washing-up liquid

ketchup

a What to keep the same to make your test fair:

_____

_____

e Outline of graph:

How does the thickness of a liquid affect how fast it freezes?

b Prediction using scientific knowledge:

_____

_____

_____

_____

d Table of results:

c Relevant observations to make:

_____

_____

_____

_____

Investigate freezing

1 You will plan an investigation to answer Jin's question.

Does water change in volume when it freezes?

a Predict the answer to Jin's question. Explain your thinking.

_____

_____

_____

b What will you do in your investigation? Draw a labelled diagram.

c What will you observe or measure?

_____

d How will you make sure that your observations or measurements are reliable?

_____

_____

e What problems do you think you might have when you carry out the investigation?

_____

_____

_____

f What could you do to avoid these problems?

_____

_____

_____

Thermal conductivity cartoon

1 Create a cartoon to explain what happens if one ice cube is placed on a plastic spoon, and another ice cube is placed on a metal spoon at the same time.

Decide which facts to include, how to present them as a cartoon, and the best order for them.

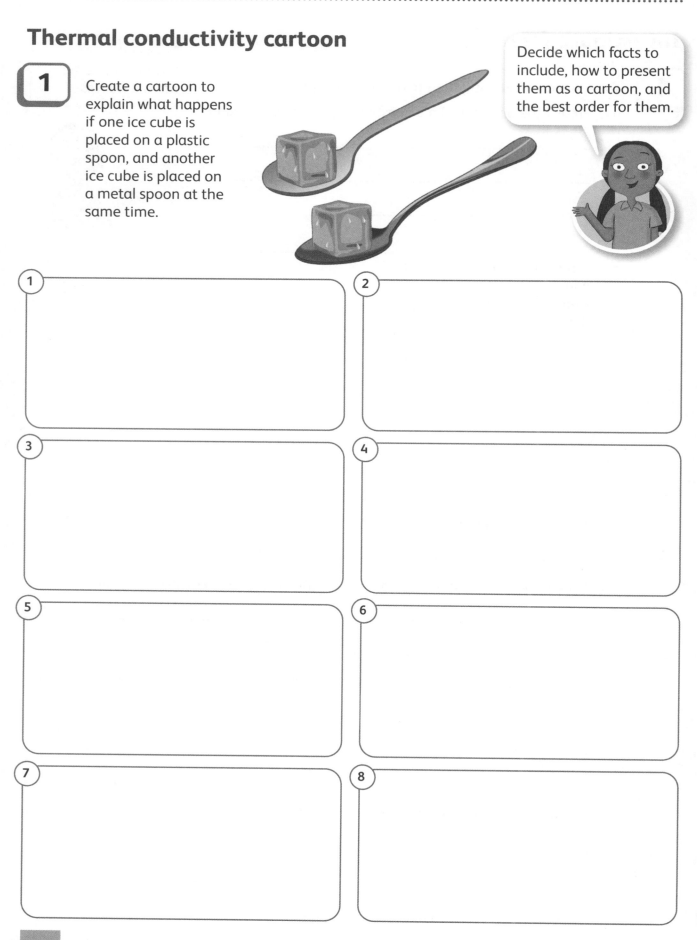

1	2
3	4
5	6
7	8

Boiling and evaporation

1 **a** What do you need to do to water to make it boil?

_____

  b What are the bubbles in boiling water made from? _____

  c What is the steam that comes from boiling water made from?

_____

  d What is the boiling point of water? _____

  e Is it possible to heat water above its boiling point? Explain your answer.

_____

_____

  f How does the boiling point of saltwater compare to the boiling point of freshwater?

_____

  g Can evaporation occur on a cold day? Explain your answer. _____

_____

_____

2 **a** Tick true (✔) or false (✗) for each statement below.

Statement	True	False
All liquids boil at the same temperature.		
When a liquid boils, it turns into a gas.		
When a liquid evaporates, it turns into a gas.		
Liquids can only turn into a gas at boiling point.		
Water vapour is a liquid.		
Steam is a gas.		

Irreversible changes, product and reactants

1 Circle the examples below that are irreversible changes.

concrete being mixed

ice lolly melting

dew forming on the grass

brown sugar stirred into water

cake baked in an oven

water droplets forming on a window

chocolate left in the Sun

a fried egg in a pan

2 Complete the table by writing the reactants or products for each irreversible change. One has been done for you.

Reactants	Product
butter, sugar, eggs, flour	cake
bicarbonate of soda, citric acid, olive oil, essential oil, food colouring, dried flower petals	
	fried egg
cement, sand, water, limestone	

3 Explain how we can tell if an irreversible change has taken place.

_____

_____

_____

Material changes crossword

1 Use the clues below to complete this crossword.

Across

2 This sort of change has taken place if a material can be changed back to the way it was before.

6 The process in which a material changes state from a gas to a liquid.

7 What everything you see around you is made of.

8 The process in which a liquid changes state to become a solid.

9 The opposite of 8 across.

Down

1 The opposite of 2 across.

3 A puddle drying up is an example of this change of state.

4 The temperature at which a solid melts is called its melting _____.

5 A liquid does this when evaporation takes place all the way through the liquid, not just on the surface.

Changes during cooking

1 Some Class 6 learners observed changes in everyday substances.

Complete their table of results.

Change	What causes the change (heating, mixing, both, or something else)?	Is the change reversible or irreversible?	How do you know if the change is reversible or irreversible?
a freezing orange juice to make ice lollies			
b toasting marshmallows			
c frying an egg			
d melting cheese			
e baking biscuits			
f browning onions			
g putting butter in the refrigerator to firm up			
h cooking corn kernels so they burst to make popcorn			

Unit 4 Reversible and irreversible changes

Self-check

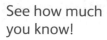
See how much you know!

 I can do this.

 I can do this, but I need to keep trying.

 I can't do this yet.

What can I do?	😄	😐	🙁
1 I can recognise and describe the processes of condensation, evaporation, melting, freezing, boiling and dissolving.			
2 I can explain that the temperature at which a material changes state varies for different substances.			
3 I can name three materials that are good thermal conductors and three that are not good thermal conductors.			
4 I can link freezing, melting, evaporation and condensation to the changes of state of a substance.			
5 I can describe and explain the differences between evaporation and boiling.			
6 I can state what reactants and products are.			

I need more help with:

_____

_____

_____

Force diagrams

Arrows are used in force diagrams to help us understand the forces acting on objects that are interacting on each other. Some Class 6 learners made a concept map to show everything they know about forces and force arrows. The group is not confident that they have understood everything and think they may have made some mistakes.

1 Check the concept map to see if it contains any errors.

2 Circle the errors you find and correct them on the concept map.

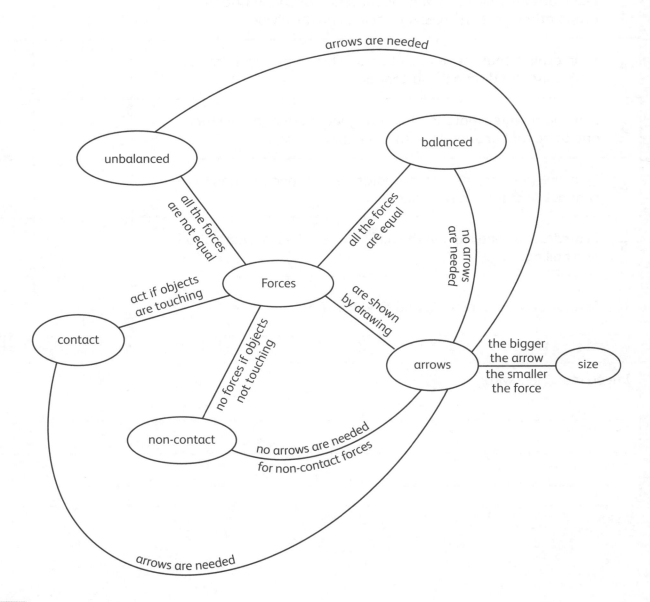

Balanced and unbalanced forces

1 Some of the diagrams below show balanced forces; others show unbalanced forces. The force acting on each object affects the way it moves. Write the letter of each force diagram (**a**, **b**, **c** and **d**) in the correct part of the table below.

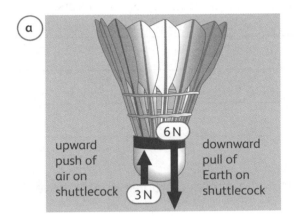

a
6 N
upward push of air on shuttlecock
3 N
downward pull of Earth on shuttlecock

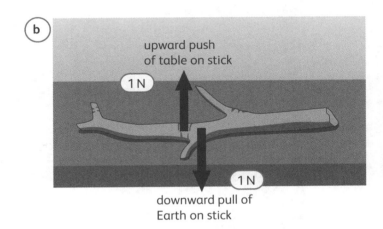

b
upward push of table on stick
1 N
1 N
downward pull of Earth on stick

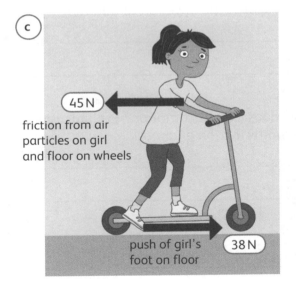

c
45 N
friction from air particles on girl and floor on wheels
push of girl's foot on floor
38 N

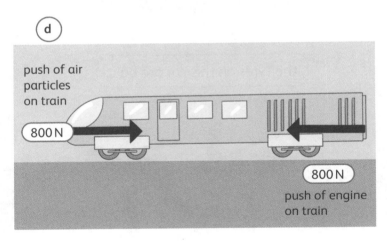

d
push of air particles on train
800 N
800 N
push of engine on train

Motion	Forces	
	Balanced forces	**Unbalanced forces**
not moving		
constant speed		
speeding up		
slowing down		

Friction: the force that opposes motion

1 Look at the pictures and answer the questions.

a

high friction between a car's tyres and the road

Explain, using the words *friction* and *motion*, why high friction is good for car drivers.

_____

_____

_____

_____

b

the tyres on this car are bald

Why are bald tyres a problem for the driver?

_____

_____

What should the driver do about it?

_____

_____

c

low friction between the skater's skates and the ice

How does ice affect motion? Explain, using the words *friction* and *motion*.

_____

_____

_____

_____

d

low friction between the girl's socks and the polished floor

Is low friction in this situation helpful

or is it a problem? _____

Why? _____

_____

_____

_____

Air resistance

1 These three cars have the same mass. They have identical wheels and an identical engine. They all travel the same distance at the same speed.

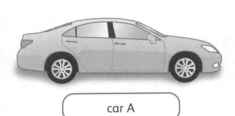
car A

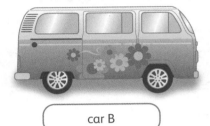
car B

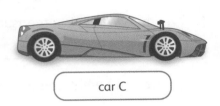
car C

  a Which car will use the least fuel during the journey? _____

  b Explain the answer you gave to question **a**. Use these scientific words:

  (surface area)   (air resistance)   (direction)

  _____

  _____

  _____

2 Look at these positions a skydiver can take.

position A

position B

position C

  a In which position will the skydiver reach a slower final speed? _____

  b Explain the answer you gave to question **a**. Use these scientific words:

  (surface area)   (air resistance)   (direction)

  _____

  _____

  _____

Mass and weight

1 Answer these questions about mass and weight.

a Which downward force is exerted by an object that is being pulled by gravity:

mass or weight? _____

b Which is a measure of the amount of matter an object contains: mass or weight?

_____

c In what unit is weight measured? _____

d In what unit is mass measured? _____

e True or false: If one object has more mass than another, it also weighs more. _____

f How many newtons does an object with 1 kilogram of mass weigh? _____

2 On Earth, an astronaut has a mass of 90 kilograms and a weight of 900 newtons. Now imagine the same astronaut floating in space, far away from the pull of Earth's gravity.

a What is the astronaut's mass in space? Explain your thinking.

_____

_____

_____

_____

_____

_____

_____

b What is the astronaut's weight in space? Explain your thinking.

_____

_____

_____

_____

_____

Leaving the Moon

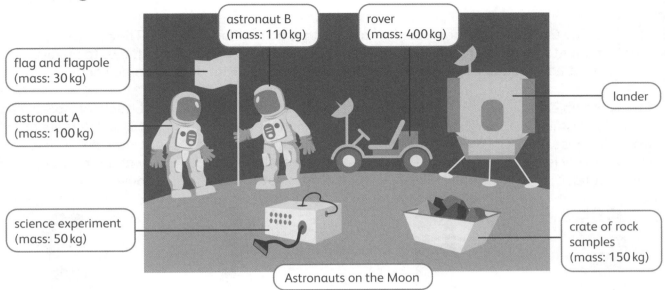

flag and flagpole (mass: 30 kg)

astronaut B (mass: 110 kg)

rover (mass: 400 kg)

lander

astronaut A (mass: 100 kg)

science experiment (mass: 50 kg)

crate of rock samples (mass: 150 kg)

Astronauts on the Moon

1 It is time for the astronauts to leave the Moon in the lander. First, they must find out if the lander has enough fuel to carry all the objects they want to take back to the spacecraft.

a The lander has 200 litres (ℓ) of fuel left. It will use 1 ℓ of fuel for every 3 newtons (N) it carries back. What is the maximum weight the lander can carry back to the spacecraft? _____N

b Help the astronauts to work out the weight of each person and object on the Moon by completing the table below. The first row has been done for you.

weight (N) = mass (kg) × strength of gravity (N/kg)

Person/Object	Mass (kg)	Strength of gravity on the Moon (N/kg)	Weight of object on the Moon (N)
astronaut A	100 kg	1.6 N/kg	160 N
astronaut B		1.6 N/kg	
rover		1.6 N/kg	
crate of rock samples		1.6 N/kg	
science experiment		1.6 N/kg	
flag and flagpole		1.6 N/kg	
Total weight on the Moon			

c Does the lander have enough fuel to carry everything back to the spacecraft? _____

d If you answered 'No' to question **c**, which objects should the astronauts leave on the Moon? Explain your thinking.

_____

_____

_____

Weight in water

Some Class 6 learners collected six objects that do not float in water. They used a force meter to weigh each object in air. Then they used their knowledge of forces to predict what each object would weigh in water. They came up with different predictions.

I think the object will weigh less in water, because the pull of gravity is less in water.

I think the object will weigh less in water, because the water will push up on the object.

I think the object will weigh the same in water, because the size of the object does not change.

I think the object will weigh more in water, because the water is pushing down on top of it.

Anya

Diya

Henna

Priya

The learners used the force meter to weigh the objects in water. They collected this data:

a Use the data to complete this table of results.

Object	A	B	C	D	E	F
Weight in air (N)						
Weight in water (N)						

Weight in air:

object **A** 25 N object **D** 10 N object **F** 34 N

object **B** 19 N object **C** 9 N object **D** 28 N

Weight in water:

object **D** 3 N object **C** 2 N object **A** 5 N

object **F** 9 N object **E** 7 N object **B** 4 N

b Use the table of results to draw a bar chart.
Draw two bars for each object.

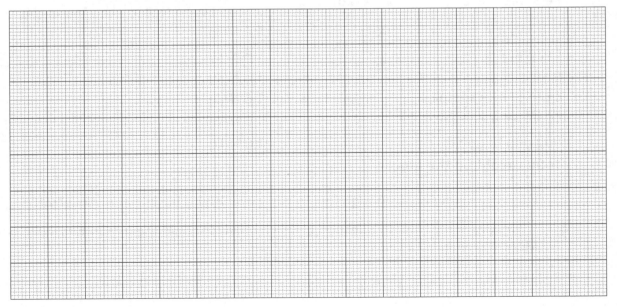

c Which Class 6 learner/s made a prediction that matched the results? _____

d Which Class 6 learner/s had the correct science thinking to explain the results? _____

Unit 5   Forces

Self-check

See how much you know!

 I can do this.

 I can do this, but I need to keep trying.

 I can't do this yet.

What can I do?	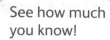		
1 I can use force diagrams to show the name, sizes and directions of forces acting on objects.			
2 I can describe the effect of forces on an object at rest.			
3 I can describe the effect of forces on an object that is moving.			
4 I can describe the difference between mass, measured in kilograms (kg), and weight, measured in newtons (N).			
5 I can explain how the weight of an object changes, but the mass does not.			
6 I can link ideas about mass and weight to the gravitational attraction between masses.			
7 I can describe how friction, including air resistance and water resistance, affects the way objects move.			
8 I can understand that every object has a mass, including gases.			

I need more help with:

_____

_____

_____

Electricity mind map

1 Create a mind map to show what you know about electricity. Use some of these words:

electrical appliance   safety   electric shock   mains electricity

cell   components   circuit   electric current   switch

> electricity

Keep adding to this mind map throughout the unit. When you learn something new, use different-coloured pencils or pens to add to it.

Circuits

1 Write instructions for how to make a circuit that will make a buzzer sound. Include a diagram of the circuit. Label each component.

2 On a sheet of paper, write instructions for how to make a circuit that will make a lamp light. Include a diagram of the circuit. Label each component.

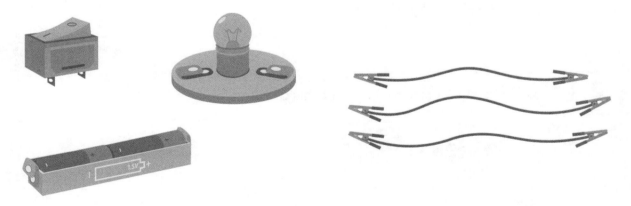

Give three possible reasons why the lamp did not light.

- _____
- _____
- _____

Circuit diagrams

1 List the components needed for each of these circuit diagrams.
There is one example for you.

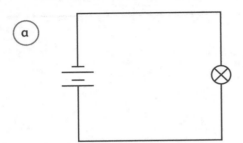

Components:

two wires, two cells, one lamp

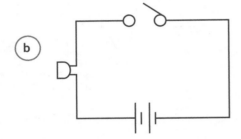

Components:

_____

_____

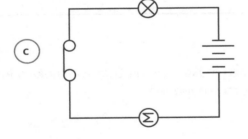

Components:

_____

_____

2 Draw a circuit diagram to represent this circuit.

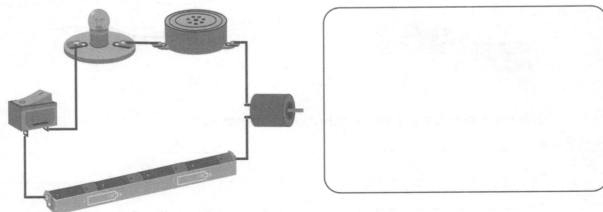

Electrical cells

1 Do research to answer these questions about electrical cells.

  a Who invented the first electrical cell?

  _____

  When? _____

  What was it called? _____

  b How does an electrical cell work?

  _____
  _____
  _____
  _____

  c What are the dangers of electrical cells?

  _____
  _____
  _____
  _____

  d What should you do with electrical cells and batteries that no longer work?

  _____
  _____
  _____
  _____

2 **a** Ask your own question about electrical cells.

  _____
  _____

  b Now answer your question.

  _____
  _____

Electrical circuits quiz

1

a   What is the general name given to electrical devices connected together in a circuit?

_____

b   What is the name of the circuit when electrical devices are connected:

   i   in the same loop? _____

   ii   in different loops? _____

c   What does this circuit symbol represent?

_____

d   What is wrong with this circuit diagram?

_____

_____

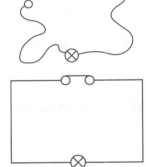

e   What is wrong with this circuit diagram?

_____

_____

2   Write your own quiz questions about circuits.

a   _____

_____

b   _____

_____

c   _____

_____

d   _____

_____

e   _____

3   Give the questions you wrote in question 2 to a partner to answer.

How many questions did your partner answer correctly?

Match circuit component names to their symbols

1 Write the correct name of the circuit component represented by each symbol.
Choose from the names of the components below:

cell   wire   open switch   buzzer   lamp

battery   junction of conductors   closed switch

Series circuits and parallel circuits

1 Which circuits below are series circuits? Which are parallel circuits?
Write *series* or *parallel* underneath each circuit diagram.

a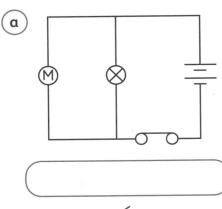

b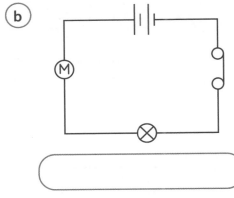

c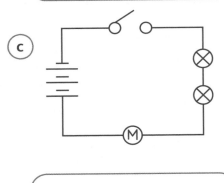

d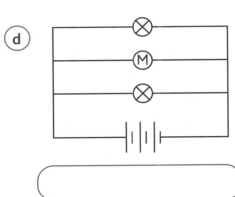

2 This is a parallel circuit.

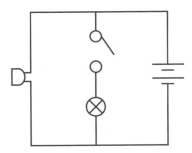

Draw a circuit diagram that shows the same components in a series circuit.

3 Look at the diagram of a parallel circuit and your drawing of a series circuit in question 2.
When might you choose to use a parallel circuit rather than a series circuit?

_____

_____

Harry's switches

1 Harry made a model house with a light in every room.

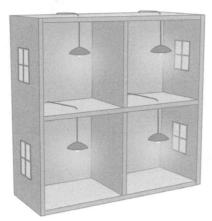

This diagram shows the circuit in the model house.

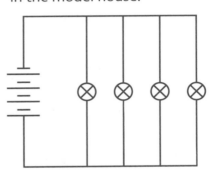

a Harry wants to be able to switch each light in the house on and off separately. Draw a cross (✗) on the diagram above to show where he should put each switch.

b Harry also wants to be able to switch all the lights in the house on and off at the same time. Draw a circle on the diagram above to show where he should put the switch to do this.

c Harry does not have any switches. He only has these materials shown here.

Write a set of instructions for Harry on how to make a switch. Include a labelled diagram which uses the symbol for a switch. Say whether the diagram shows the switch on or off.

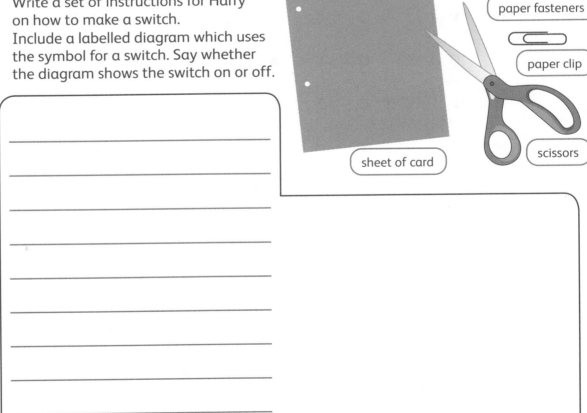

paper fasteners

paper clip

scissors

sheet of card

_____

_____

_____

_____

_____

_____

_____

_____

Electrical word search

1 Find and circle these words in the word search below:

| conductor | insulator | parallel | series | circuit |

| electricity | voltage | volts | components | lamps |

The words may be written in any of these directions:

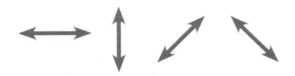

C	I	R	C	U	I	T	A	R	T	U	I
H	E	O	O	B	E	S	Q	I	R	Y	S
G	C	I	N	S	U	L	A	T	O	R	S
H	V	I	D	C	P	L	C	H	B	E	T
E	O	A	U	K	A	F	P	K	R	A	N
E	L	E	C	T	R	I	C	I	T	Y	E
N	T	H	T	O	A	S	E	O	O	A	N
F	A	D	O	I	L	S	P	A	J	A	O
T	G	P	R	G	L	B	L	M	G	P	P
A	E	E	M	F	E	T	D	D	A	L	M
L	O	J	S	T	L	O	V	O	E	L	O
L	E	T	I	M	E	S	K	G	S	V	C

Self-check

See how much you know!

 I can do this.

 I can do this, but I need to keep trying.

 I can't do this yet.

What can I do?			
1 I can draw a diagram of an electrical circuit using circuit symbols.			
2 I can use diagrams and conventional symbols to draw series and parallel circuit diagrams.			
3 I can build series and parallel circuits.			
4 I can predict and test what happens when components are added to a series or parallel circuit.			
5 I can create electrical devices for different jobs using my ideas about electrical circuits.			

I need more help with:

_____

_____

_____

Light and seeing

1 Tick (✔) the diagram that shows how we see things.

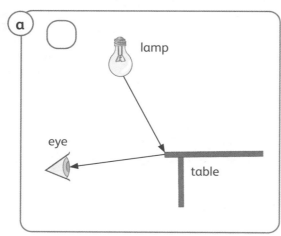

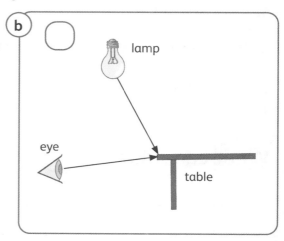

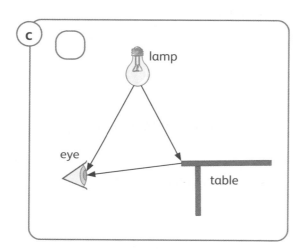

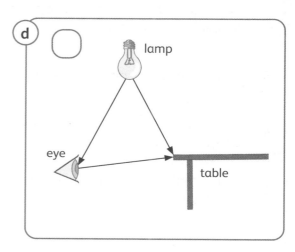

2 Draw light rays on this picture to show how Theresa sees the butterfly.

Travelling light

1 Three Class 6 learners wrote sentences about how light travels, using these words:

> arrow   beam   ray diagram   rays   spreads out   straight

Read what each one wrote.

A Light travels as a beam. Light travels in straight lines and spreads out. We use arrows in a ray diagram to show where the light rays are.

B Luminous objects give off light. Light travels in straight lines as it leaves the light source. As the light travels away from its source, it spreads out. We cannot see the light, so to help us understand what is happening, we represent the light by drawing straight lines. The straight lines are called rays. To show the direction of the light and where it goes, we use arrows in a ray diagram.

C Beams of straight arrow lines called rays are in a ray diagram. These rays can sometimes spread out.

a Which paragraph do you think is best? Which is worst? Rank them in order, from best to worst. Explain why you have put them in this order.

Order:	
Best _____ **Worst**	
Reasons:	

2 What will happen when each light source is switched on? Draw a ray diagram to show how the light from each light source will travel.

Picture   **Ray diagram**

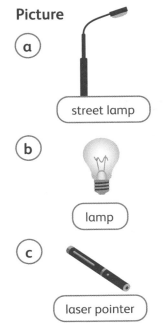

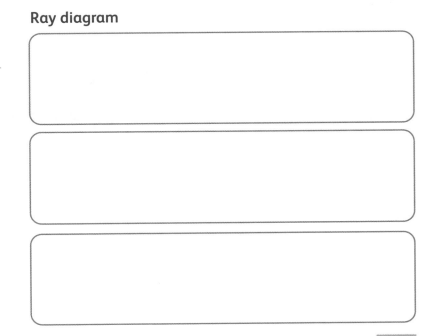

(a) street lamp

(b) lamp

(c) laser pointer

Reflections

1   **a**   Which diagram below shows a specular reflection, and which a diffuse reflection?

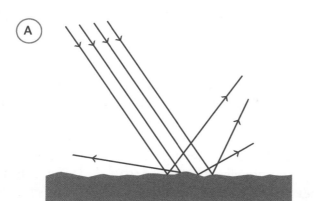

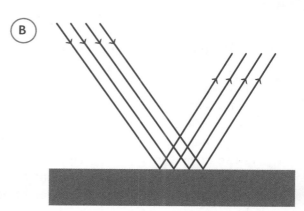

Specular reflection: _____. Explain your reasons: _____

Diffuse reflection: _____. Explain your reasons: _____

Give an everyday example of specular reflection: _____

Give an everyday example of diffuse reflection: _____

b   Add these labels to the diagrams above:

| very smooth surface | incoming rays of light | uneven surface | reflected rays |

2   A pencil is reflected on the surface of a metal spoon.

a   How is the reflection of the pencil similar to the real pencil?

_____

_____

b   How is the reflection of the pencil different to the real pencil?

_____

_____

Incoming and reflected rays

1 Measure angle *a*, at which each light ray hits the mirror.
Write the size of the reflected light ray next to each diagram.

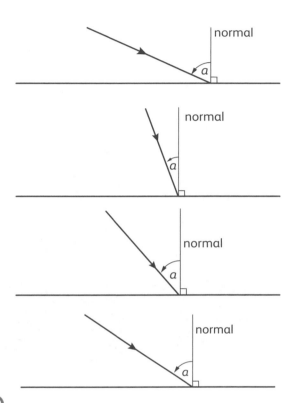

a _____

b _____

c _____

d _____

2 Measure angle *b*, at which each light ray is reflected from the mirror.
Write the size of the incoming light ray next to each diagram.

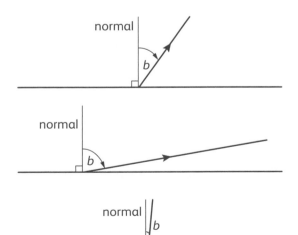

a _____

b _____

c _____

d _____

What have you learnt so far?

1 Look at the picture and answer the questions.

a Circle the three light sources in the picture.

b Order the light sources from brightest to dimmest by adding the numbers 1, 2 or 3 to the picture. The brightest light source is 1. The dimmest is 3.

c Describe three ways in which the people in the picture are using light.

_____

_____

_____

d Draw rays on the picture to show how the light from the table lamp is travelling.

e Colour in the T-shirt of the person who experiences the brightest light from the table lamp.

f Draw a circle in the picture around an object that is producing a reflection.

g Look at the reflection in the picture. How is the reflected image different to the object being reflected?

_____

_____

_____

Mirror maze

1 Class 6 learners made a maze using mirrors. They shone a laser pointer onto the first mirror in the maze.

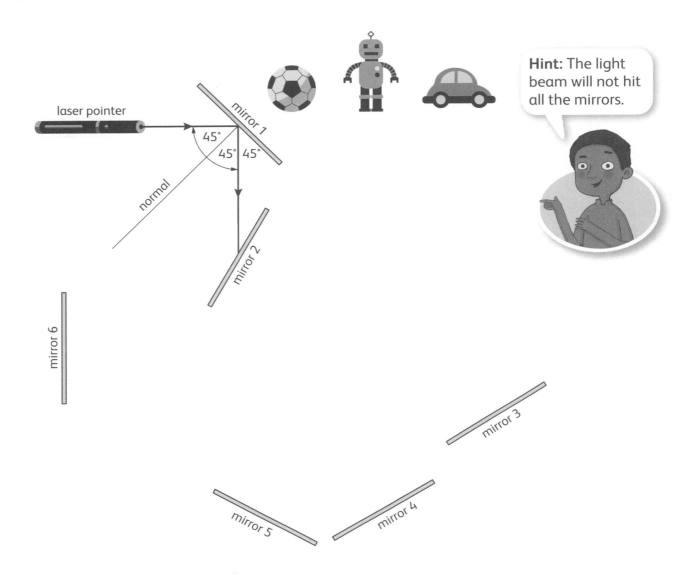

a Circle the object that you think the light from the laser pointer will reach.

b Draw the path that the light beam will take. As you do this, write the angle at which the light beam hits each mirror and the angle at which it is reflected. The angles for the first mirror have been done for you.

c Was your prediction in question **a** correct? _____

2 Draw your own light maze on a sheet of plain paper. Give it to a partner to solve.

Light meeting transparent objects: refraction

1 Angle *a* is the angle at which a light ray hits the surface of a transparent object. Is the angle of refraction (angle *b*) larger, smaller or the same as angle *a*? Explain your reasons.

One has been done for you.

a

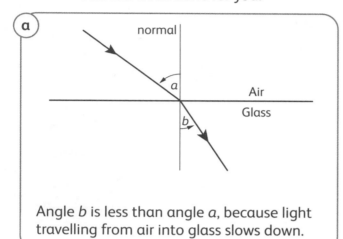

Angle *b* is less than angle *a*, because light travelling from air into glass slows down.

b

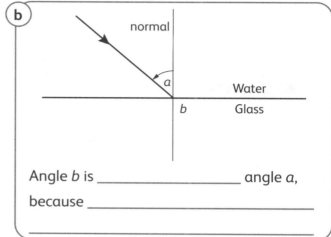

Angle *b* is _____ angle *a*,

because _____

_____

c

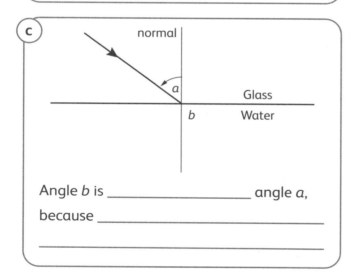

Angle *b* is _____ angle *a*,

because _____

_____

d

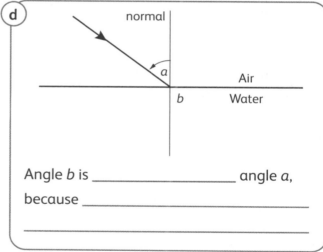

Angle *b* is _____ angle *a*,

because _____

_____

Now make up your own diagram to test on a partner.

e

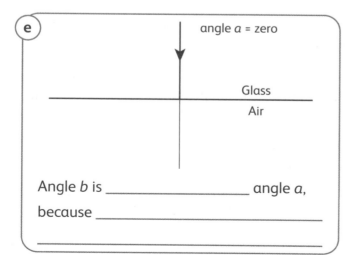

Angle *b* is _____ angle *a*,

because _____

_____

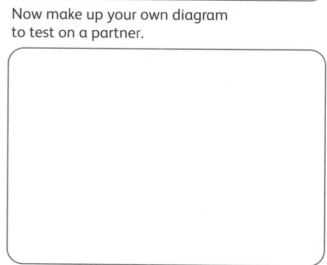

Unit 7   Light, reflection and refraction

Self-check

 I can do this.

 I can do this, but I need to keep trying.

 I can't do this yet.

See how much you know!

What can I do?	😄	😐	🙁
1 I can draw a ray diagram to represent how light travels.			
2 I can state what happens when light meets an opaque object.			
3 I can describe how a ray of light changes direction when it is reflected from a mirror.			
4 I can state what happens when light meets a transparent object.			
5 I can describe how the direction of a light ray will change when it travels through different transparent materials.			
6 I can explain why the direction of a light ray changes.			
7 I can investigate different everyday occurrences of refraction.			

I need more help with:

_____

_____

_____

The structure of the Earth

1 **a** Label the parts A, B, C and D. Use these words to help you:

(inner core) (outer core) (mantle) (crust)

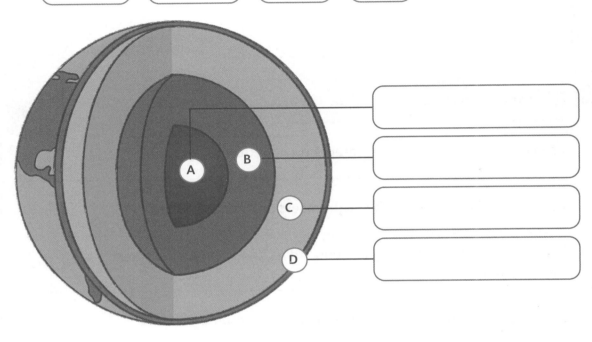

b Which parts of the Earth are made of solid rock? _____

c Which parts of the Earth are made of molten (liquid) rock? _____

d What is the name given to molten rock when it is below the surface of the Earth?

_____

e What is the name given to molten rock when it comes out onto the surface

of the Earth? _____

2 Molten rock can escape through openings in the Earth's surface through volcanoes.

If the answer is *volcano*, what is the question?

Write three questions that could be answered with the word *volcano*.

a _____

b _____

c _____

Different types of sedimentary, metamorphic and igneous rocks

 Rocks that let water through them are called *permeable*. Rocks that do not let water through them are called *impermeable*.

Guss was given microscope slides of five different types of rocks to look at.

sandstone (sedimentary rock) particles seen under a microscope

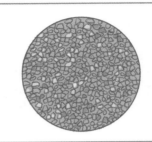 

marble (metamorphic rock) particles seen under a microscope

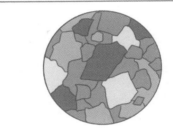 

limestone (sedimentary rock) particles seen under a microscope

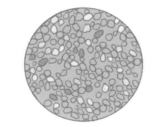 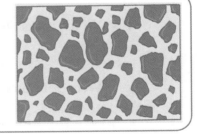

chalk (sedimentary rock) particles seen under a microscope

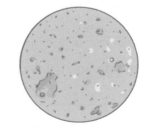 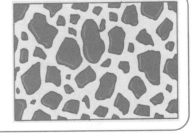

granite (igneous rock) particles seen under a microscope

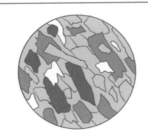 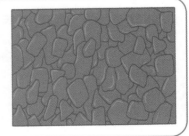

Guss thinks that soft sedimentary rocks have more holes in them than metamorphic and igneous rocks. Guss looked at the slides, and made a prediction about which rocks are permeable and which are impermeable.

a   Which rocks do you think Guss predicted were permeable? _____

Why? _____

b   Which rocks do you think Guss predicted were impermeable? _____

Why? _____

2   Guss tested some rocks to see if they let water through. Here are his results:

Rock	Did it let water through?
sandstone	yes
marble	no
limestone	yes
chalk	yes
granite	no

a   Which rocks let water through?

_____

b   Which rocks did not let water through?

_____

c   Were your predictions for question 1 correct?

_____

d   Why would it be a bad idea to make a house out of chalk?

_____

e   Which of the rocks tested do you think should be used for each of the following?

Making footpaths _____

Why? _____

Marking gravestones _____

Why? _____

Is it metamorphic, igneous or sedimentary?

1 Makrana marble is a metamorphic rock made of white marble. It is used in sculpture and in buildings. It is mined in the town of Makrana in Rajasthan, India, and was used in the construction of the Victoria Memorial in Kolkata. Tick (✔) all the words for objects below that you think could be made of marble:

(stairs) (tables) (carpets) (shirts)

(floor tiles) (statues) (cushions)

Victoria Memorial in Kolkata

2 Use your research skills to answer these questions about these rocks: limestone, slate, granite.

a What is the colour of each of the rocks?

_____

_____

b Give an example of what each rock can be used for.

_____

_____

c Which rock is sedimentary, igneous or metamorphic?

_____

_____

3 Carry out some research.

a What kind of rock is pumice?

_____

b What is special about pumice?

_____

_____

c What is pumice used for?

_____

_____

d Draw a picture of pumice.

Different types of rocks

1 There are three different types of rocks: sedimentary, metamorphic and igneous. Use your research skills to identify each type of rock in the table and some of its properties. Tick (✔) the correct columns. One has been done for you.

Rock	Sedimentary	Metamorphic	Igneous	Crystals	Rock fragments	Fossils
slate		✔		✔		
sandstone						
granite						
marble						
limestone						
basalt						
chalk						

2 Fossils are found in some rocks. Pia made this fact file about her favourite fossil.

Name: trilobite

Lived: on the ocean floor

Last alive: 250 million years ago

Main food: sea worms

Complete a fact file about your favourite fossil. Choose your own headings.

You might have to do some research first!

Mind map of processes in the rock cycle

1 Create a mind map, using the words below along with your own notes, to show your understanding of the processes in the rock cycle.

solidification   erosion   sedimentation   burial   metamorphism

The rock cycle

1 Over time, rocks of one type can change into rocks of another type. This is called the rock cycle. For each of the examples below, draw what happened in the rock cycle long before, before, after, and long after the image.

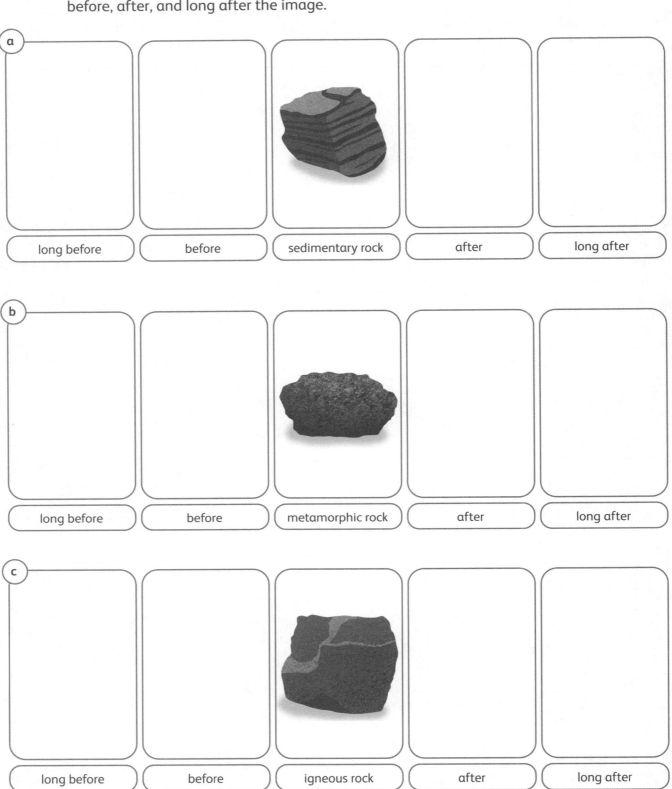

a

| long before | before | sedimentary rock | after | long after |

b

| long before | before | metamorphic rock | after | long after |

c

| long before | before | igneous rock | after | long after |

Soil

Some Class 6 learners made this edible model of soil.

organic material

topsoil

subsoil

bedrock

They used different edible materials to represent the different layers:
- crushed biscuits
- chocolate and butterscotch chips
- shredded coconut coloured green
- gummy worms
- chocolate mousse.

a   Complete the table below of how the edible layers represent soil layers.

b   Do you think the model they made is a good model? For each of the different layers, explain the strengths and the weaknesses of the model.

Layer	Materials used	Strengths of the model	Weaknesses of the model
organic material			
topsoil			
subsoil			
bedrock			

The Class 6 learners then tested three different soils for drainage: sandy soil, rocky soil, and clay soil. They poured 100 cm³ of water through each soil and made a note of the amount of water they collected. They presented their results in a table.

Soil sample	Volume of water poured in (cm³)	Volume of water collected (cm³)	Amount of water held by the soil
A	100 cm³	40 cm³	
B	100 cm³	95 cm³	
C	100 cm³	80 cm³	

a   Work out how much water each soil retained (held onto). Fill in the final column of the table.

b   Which soil do you think is the sandy soil, the rocky soil and the clay soil?

A _____ Why? _____

B _____ Why? _____

C _____ Why? _____

c   Explain why water flows through different soils at different rates.

_____

_____

Rocks in a house

1 Look at the picture of the house where Rania lives.

Complete the labels by writing the name of the rock or soil material used to make each part. One has been done for you. Use these words to help you:

marble   sandstone   slate   clay   granite

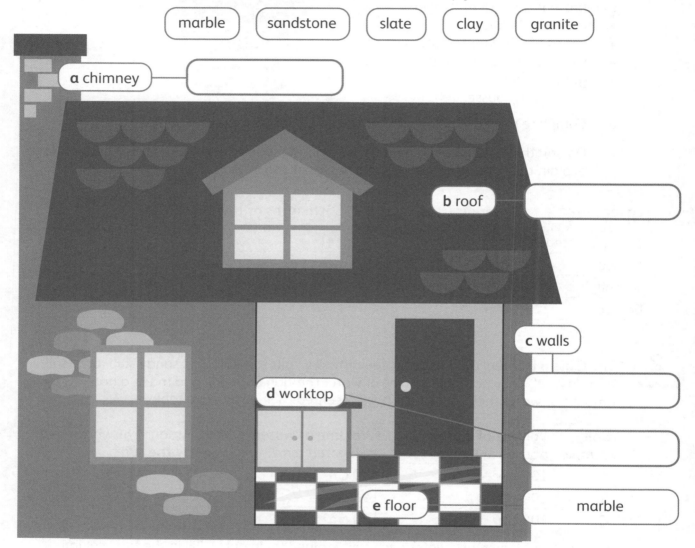

a chimney

b roof

c walls

d worktop

e floor   marble

2 **a** In column 1, on the table below, write the names of four other objects inside and outside your home that are made from a type of rock.

b In column 2, identify which of the three types of rock it is made from.

c Explain the advantages of using this type of rock for the job.

Name of object	Type of rock it is made from	Advantages of using this type of rock

Self-check

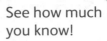
See how much you know!

 I can do this.

 I can do this, but I need to keep trying.

 I can't do this yet.

What can I do?			
1 I can explain how the different features of rocks make them useful to us.			
2 I can describe in which type of rocks fossils are formed.			
3 I can explain how fossils are made.			
4 I can describe the rock cycle.			
5 I can recognise and describe how the processes of solidification, erosion, sedimentation, burial, metamorphism and melting are linked to rocks.			
6 I can explain how one type of rock can change into another.			
7 I can name the different types of soil.			
8 I can name the different layers of soil.			
9 I can explain why water flows through different types of soil at different rates.			
10 I can investigate how plants help maintain soil quality.			

I need more help with:

_____

_____

_____

Unit 9　Earth and the Solar System

Earth, Sun and Moon

1 Label the Sun, the Earth and the Moon in the diagram below.

2　**a** Use your research skills to find out three interesting facts about the Earth, the Sun and the Moon.

Three interesting facts about the Earth	Three interesting facts about the Sun	Three interesting facts about the Moon

b Share what you have found out with a partner.

The moving Earth

1 **a** Draw a diagram below to show how the Earth moves.

b What is the name of the imaginary line around which the Earth rotates?

_____

c How long does it take the Earth to complete one rotation?

_____

d Why did ancient people think the Sun travelled around the Earth?

_____

e What is the name of the force that keeps the Earth in orbit around the Sun?

_____

f How long does it take the Earth to complete one orbit around the Sun?

_____

2 **a** Write three more questions below about the Earth and how it moves.
Include the answers to the questions.

- _____

_____

- _____

_____

- _____

_____

b Read out your questions to a partner. How many questions did they answer correctly?

_____

Objects in the Solar System

1 a   Draw a line to match the scientific word with its definition.

Scientific word   **Definition**

planet

the name given to a luminous object in the sky that is made of gas and gives out light

star

the name given to a rocky or gas object that is round, orbits around a star, and is the only object in its orbit

satellite

the name given to an object that orbits around another

b   For each scientific word, provide a named example in our Solar System:

• planet: _____

• star: _____

• satellite: _____

2 Unscramble the names of the planets in our Solar System. Put them in the correct order, starting with the one closest to the Sun.

rams   mycurer   heart   sunve   sunart

truejip   utenpen   sunaru

From the Sun:

1 _____

2 _____

3 _____

4 _____

5 _____

6 _____

7 _____

8 _____

Planetary awards

1 Imagine that there is an award for the planets of the Solar System. Which of the planets would win each of these awards?

Use your research skills to answer these questions.

a

greatest number of moons

b

strongest winds

c

strongest magnetic field

d

shortest year

e

longest day

f

largest volcano

g
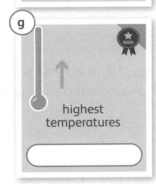
highest temperatures

h

shortest day

i

longest year

2 Invent your own planetary award and draw the certificate in the space. On your certificate, write:

- what you are giving the award for

- the name of the planet that has won the award.

Planet distances

This table shows the distance of each planet from the Sun in millions of kilometres (km).

a Fill in the last column of the table. The first two have been done for you.

Planet	Average distance from the Sun (millions of km)	Distance between the orbits of … (millions of km)
Mercury	57	
Venus	108	Mercury and Venus: 51
Earth	150	Venus and Earth: 42
Mars	228	Earth and Mars:
Jupiter	779	Mars and Jupiter:
Saturn	1 430	Jupiter and Saturn:
Uranus	2 880	Saturn and Uranus:
Neptune	4 500	Uranus and Neptune:

b What patterns do you notice in the data in the table?

_____

_____

c Imagine that there is another planet, further away than Neptune. Predict its average distance from the Sun. Use the patterns you noticed in question **b** to help you.

_____

_____

d In 1596, the astronomer Johannes Kepler predicted that there was a planet between Mars and Jupiter, because he thought the gap between the orbits of Mars and Jupiter was too big. At what distance from the Sun do you think the mystery planet may orbit? Explain your reasons.

_____

_____

e Use your research skills to find out what astronomers eventually discovered between Mars and Jupiter.

_____

_____

Moon phases

1 Look at the phases of the Moon. Do some research to find out the names of the stages.

   a Order the phases by writing the numbers 1 to 8 in the boxes.

   b Write one of these labels below each picture:

(first quarter) (full moon) (new moon) (last quarter) (waning crescent)

(waning gibbous) (waxing crescent) (waxing gibbous)

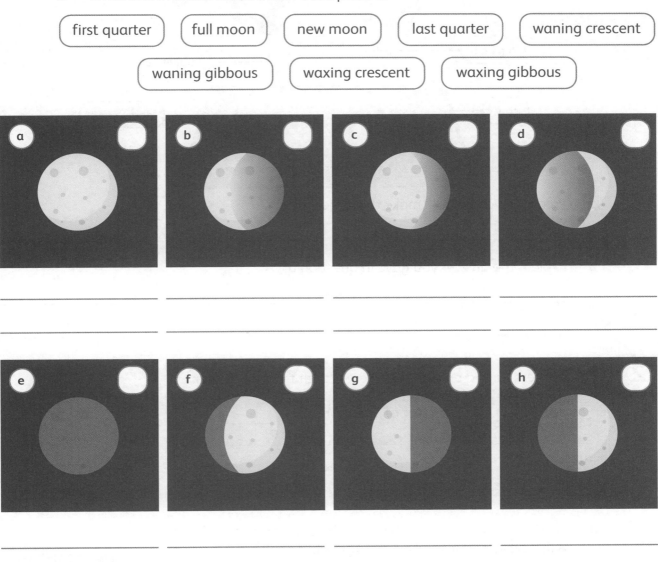

2   **a** How long does it take the Moon to complete one orbit around the Earth?

      b How do you know?

Craters on the Moon

Some Class 6 learners wanted to model making craters on the Moon by dropping balls into damp sand. They wanted to find out which ball would make a larger crater.

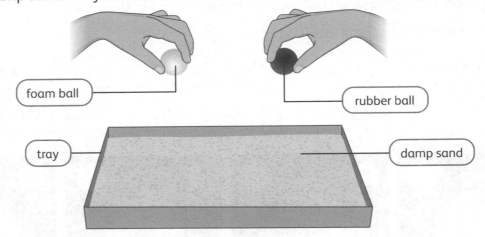

foam ball

rubber ball

tray

damp sand

a Predict which ball will make the larger crater.

_____

b Explain the answer you gave in question **a**.

_____

_____

The learners carried out the test. Here is their table of results:

Material of ball	Diameter of crater (mm)		
	Test 1	Test 2	Test 3
rubber	64	63	68
foam	45	49	42

a Predict the size of the crater made by the ball in this picture. Will it be larger or smaller than the rubber ball?

_____

b Explain the answer you gave in question **a**. Use the table of results in your explanation.

_____

_____

_____

_____

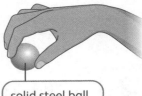

solid steel ball

Famous astronomers

1

Match each astronomer listed in the table below to the correct place and the correct contribution.

Places:
China, Greece, Italy, Persia (Iran), Poland, United States

Contributions:
* created the first heliocentric (Sun-centred) model
* created a more advanced heliocentric model
* provided evidence of other galaxies outside the Milky Way
* calculated the diameter of the Earth
* realised the Moon's brightness was caused by reflected sunlight
* observed and studied sunspots.

	Astronomer	Place	Contribution
a	Nicolaus Copernicus		
b	Zhang Heng		
c	Edwin Hubble		
d	Aristarchus		
e	Galileo Galilei		
f	Ahmad ibn Muhammad ibn Kathir al-Farghani		

2

Write the names of the astronomers from Activity 1 in date order, from the earliest to the most recent. _____

_____

_____

Unit 9   Earth and the Solar System

Self-check

 I can do this.

 I can do this, but I need to keep trying.

 I can't do this yet.

See how much you know!

What can I do?			
1 I can define what a planet is.			
2 I can define what a star is.			
3 I can name the eight planets in our Solar System in the correct order.			
4 I can define what a satellite is.			
5 I can name the Earth's satellite.			
6 I can explain why we only see one side of the Moon even though it is spinning.			
7 I can describe the changes in appearance of the Moon over its monthly cycle.			
8 I can explain how scientists throughout history have helped develop our understanding about the Earth and space.			

I need more help with:

_____

_____

_____